PLANTS with SUPERPOWERS

PLANTS CAN HIDE!

Monika Davies

Gareth Stevens
PUBLISHING

Please visit our website, www.garethstevens.com. For a free color catalog of all our high-quality books, call toll free 1-800-542-2595 or fax 1-877-542-2596.

Library of Congress Cataloging-in-Publication Data
Names: Davies, Monika, author.
Title: Plants can hide! / Monika Davies.
Description: New York : Gareth Stevens, [2023] | Series: Plants with superpowers | Includes bibliographical references and index.
Identifiers: LCCN 2022006646 (print) | LCCN 2022006647 (ebook) | ISBN 9781538278925 (set) | ISBN 9781538278932 (library binding) | ISBN 9781538278918 (paperback) | ISBN 9781538278949 (ebook)
Subjects: LCSH: Plant defenses–Juvenile literature. | Camouflage (Biology)–Juvenile literature.
Classification: LCC QK921 .D377 2023 (print) | LCC QK921 (ebook) | DDC 581.4/7–dc23/eng/20220315
LC record available at https://lccn.loc.gov/2022006646
LC ebook record available at https://lccn.loc.gov/2022006647

Published in 2023 by
Gareth Stevens Publishing
29 East 21st Street
New York, NY 10010

Editor: Monika Davies
Designer: Leslie Taylor

Portions of this work were originally authored by Devi Puri and published as *Plants that Hide*. All new material in this edition authored by Monika Davies.

Photo credits: Cover, p. 1 Eva Pieroni/Shutterstock.com; Series Art (various accents) Ozz Design, Piotr Urakau, Designer things, Maxger/Shutterstock.com; p. 5 (upper left) https://commons.wikimedia.org/wiki/File:Euphorbia_fontqueriana.JPG; p. 5 (lower left) https://commons.wikimedia.org/wiki/File:Lithops_hookeri.png; p. 5 (lower right) lessysebastian/Shutterstock.com; p. 7 Sandra Burm/Shutterstock.com; p. 9 (stinging nettle) Haluk Cigsar/Shutterstock.com; p. 9 (close-up stinging nettle) Bildagentur Zoonar GmbH/Shutterstock.com; p. 9 (dead nettle) Elena Gromadchuk/Shutterstock.com; p. 11 (mountains) mastiffliu/Shutterstock.com; p. 11 (plant) Aleksandr Naumenko/Shutterstock.com; p. 13 jgeyser/Shutterstock.com; p. 15 (map) https://www.fs.fed.us/wildflowers/beauty/mycotrophic/images/monotropsis/MOOD2.png; p. 15 (pinesap) Svitlyk/Shutterstock.com; p. 17 (trefoil) Steffen Hauser/botanikfoto/Alamy.com; p. 17 (pine cone seeds) nulinukas/Shutterstock.com; p. 17 (pine cone branch) pavel TaraSYUK/Shutterstock.com; p. 19 (wheat) oksana2010/Shutterstock.com; p. 19 (rye) Elena Koromyslova/Shutterstock.com; p. 21 (spotty leaf) Omkoi/Shutterstock.com; p. 21 (green leaf) Tonktiti/Shutterstock.com; p. 21 (moth) Jesse John/Shutterstock.com.

Printed in the United States of America

CPSIA compliance information: Batch #CSGS23: For further information, contact Gareth Stevens, New York, New York, at 1-800-542-2595.

CONTENTS

Words in the glossary appear in **bold** type the first time they are used in the text.

Hide and Seek

If an enemy is hunting you down, what's the best **strategy** to stay safe? Some would say running far away is your best bet! But another strategy is to hide. For some plants, this is how they stay out of harm's way.

Many plants can hide—and this ability to mask themselves from enemy eyes keeps them alive. Some plants can change the color of their leaves, while others look like rocks or another plant. Let's learn more about these plants that like to stay out of sight.

Camouflage is when animals or plants use colors or shapes to blend in with their surroundings. Many animals use this trick to stay hidden—and so do plants!

On the Defense

Like all living things, plants face danger. Bugs, people, and **severe** weather can put plants at **risk**. But unlike other creatures, plants can't move freely. Instead, they must use defenses to stay safe.

A defense is a way of guarding against an enemy. Some plant **species** defend themselves with thorns, spikes, or spines. Predators know not to come near these scary-looking plants. Others may give off a bad smell or appear to look dead. Some plants can even kill their enemies with poison!

Super Fact!

Poison ivy is a plant with a clever defense. The plant makes oils that hide on its leaves. If someone touches the plant, the oils get on their hands, causing a rash, or a group of red spots on the skin.

Many plants have unseen defenses. Some might be poisonous to eat or sharp to touch. It's best to never taste or touch an unfamiliar plant.

Blending In

Plants use camouflage to change their color or patterns to match their surroundings. It's the perfect way to hide! If predators can't see a plant, they can't eat it. Another plant defense is mimicry. Mimicry is when a plant makes itself look like another plant or animal. A predator might pass by if the plant **resembles** something else.

What's the difference between the two? With camouflage, plants don't want to be noticed. With mimicry, plants may be noticed, but not for what they really are. They're hiding in plain sight!

The *Lamium album* plant, or dead nettle, looks like the stinging nettle plant. Animals avoid the stinging nettle due to its painful, stinging hairs. The dead nettle uses clever mimicry to stay safe!

stinging nettle

dead nettle

Tricky Adaptations

Adaptations are changes in a type of animal or plant that make it better able to live in its surroundings. Plant defenses of all kinds are good examples of adaptations.

Camouflage may have started in plants living long ago. Plants that blended in better with their environment, or natural surroundings, survived longer than ones that stood out. The surviving plants then passed this **trait** on to new plants. Over time, plants that hide found that blending in was the best way for their kind to stay safe.

Super Fact!

The *Fritillaria delavayi* is a small plant that sometimes has green leaves and yellow flowers. But in other areas, this plant is colored a dull tan or gray and blends in with nearby rocks. This is camouflage in action!

Humans use this plant (*fritillaria delavayi*) to make cough medicine. Some think the plant adapted and now uses camouflage to hide from humans looking for it. It grows in the Hengduan mountains in China.

Living Stone

Living stones are sometimes called pebble plants. That's because these plants look like tiny rocks! Living stones grow in the hot, dry deserts of southern Africa. They are found in sandy soils and like to get lots of sun.

Living stones have thick, rounded leaves which store water. Desert animals might love to drink from the plant's water-filled leaves. However, living stones hide by looking like stones! Their gray, green, or purple-colored leaves do not seem like a good source of water to even the thirstiest predators.

Super Fact!

In 1811, a scientist named William John Burchell was in South Africa when he picked up what he thought was a small, brown stone. It wasn't! Burchell was the first to discover a living stone.

Living stones grow sideways instead of upwards. One living stone might look like several rocks lying in a row.

Master of Disguise

Sweet pinesap is a wildflower with pink, purple, or yellow flowers and a purple stem. Unlike many plants, this wildflower does not have chlorophyll. It's one of the few plants that isn't green!

Sweet pinesap gets everything it needs to live from **fungi** living on its roots. Because it isn't green, it can blend in better with its surroundings. It's able to hide from predators by looking like dead plant matter. When predators see the sweet pinesap, they cannot tell that it's actually a healthy, living plant.

Super Fact!

Most plants make their own food with sunlight, water, chlorophyll, and **carbon dioxide**. This process is called photosynthesis. Chlorophyll is a green pigment, or a substance that gives a plant or animal color.

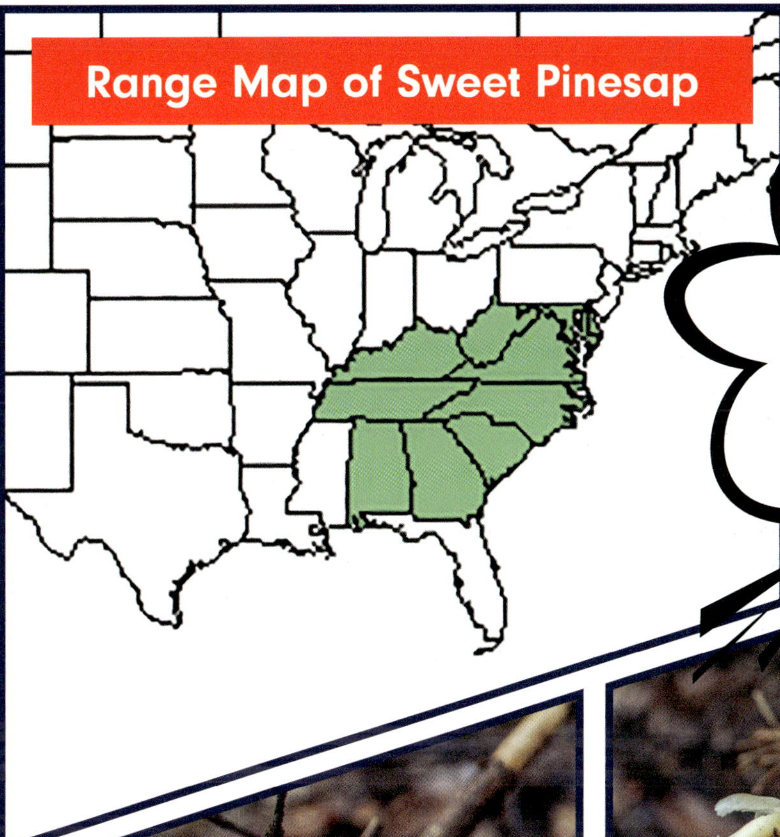

The sweet pinesap is an **endangered** plant in some areas. While it grows the most in North Carolina and Virginia, it is rare to find the plant in the wild.

Hideaway Seeds

Most plants grow from a seed. But seeds are a tasty meal for birds and other creatures. Scientists think birds are more likely to eat plant seeds that stand out in their environment. It's easier to spot a yellow or red seed in soil than a brown one. Some plants change the color of their seeds so they can hide in their native soil.

A pine tree's seeds can also hide inside pine cones. Pine cones have scales that close to hide seeds from predators!

A plant called the Chilean bird's-foot trefoil (left) produces camouflaged seeds to hide in soil too. Pine cones (right) are good at hiding their seeds inside their scales.

Wheat or Weed?

Farmers stay on the lookout for weeds in their crops. It's often easy to tell whether a plant is a weed or not. But some plants use a clever **disguise** to stay hidden.

Rye is a cereal grass that looks a lot like wheat. Today, we use its seeds as a grain, but the plant actually began as a weed. Since it looks so similar to wheat, farmers had a hard time spotting it in their fields. The overlooked rye lived and made new plants.

Super Fact!

Rye is often grown as a cover crop. Cover crops are planted to cover soil and not for food. This helps the soil stay healthy and stops **erosion**.

wheat
rye
Rye once used mimicry to hide in plain sight. But the crop is now grown in many places.

Sick Tricks

Moths are the greatest enemy of the *Caladium steudneriifolium* plant. But this plant is a master at playing pretend. Scientists believe this plant changes the color of its leaves to look like it's already been eaten. When moths see the sick-looking plant, they are less likely to approach it. And, the plant lives to see another day!

Plants use smart strategies to stay safe. Camouflage, mimicry, and other defenses help them avoid attack. How many plants do you think are hidden from the human eye?

Super Fact!

Mining moths like to lay their eggs on the caladium. Once the moth's eggs hatch, the baby caterpillars eat the plant's leaves. This is likely why the plant uses a "sick trick" to stay safe.

The caladium is an expert at faking sickness. This clever plant is found in Ecuador's rain forests.
Munch! Munch!

GLOSSARY

carbon dioxide: A gas breathed out by animals.

disguise: The state of having a false appearance.

endangered: In danger of dying out.

erosion: The act of wearing away by water, wind, or ice.

fungus: A living thing that is somewhat like a plant, but doesn't make its own food, have leaves, or have a green color. Fungi include molds and mushrooms.

resemble: To look like something else.

risk: The chance that something bad will happen.

severe: Very bad or serious.

species: A group of plants or animals that are all of the same kind.

strategy: A plan of action to achieve a goal.

trait: A feature that differentiates one thing from another.

FOR MORE INFORMATION

BOOKS

Kaner, Etta. *Pretty Tricky: The Sneaky Ways Plants Survive.* Berkeley, CA: Owlkids Books, 2020.

Leedy, Loreen and Andrew Schuerger. *Amazing Plant Powers: How Plants Fly, Fight, Hide, Hunt, & Change the World.* New York, NY: Holiday House, 2018.

WEBSITES

Adaptation
kids.britannica.com/kids/article/adaptation/352704
Discover more about adaptations in both plants and animals on this website.

Plant Defense
www.dkfindout.com/us/animals-and-nature/plants/plant-defense
Find out more about different plant defenses here.

Publisher's note to educators and parents: Our editors have carefully reviewed these websites to ensure that they are suitable for students. Many websites change frequently, however, and we cannot guarantee that a site's future contents will continue to meet our high standards of quality and educational value. Be advised that students should be closely supervised whenever they access the internet.

INDEX